Needs and Prospects for Crime Fighting Technology

The Federal Role in Assisting State and Local Law Enforcement

William Schwabe

Supported by the
Office of Science and Technology Policy

RAND
Science and Technology Policy Institute

Most policing in the United States is done by law enforcement agencies at the local level. Although most Americans prefer that policing be controlled locally, there is considerable support for federal help in funding police. One area in which federal funding has been seen as useful is in the development, testing, and implementation of improved technology. An initiative proposed by the Clinton administration would increase funding for state and local law enforcement, with emphasis on technology assistance, technology deployment, crime lab improvements, and training. This report provides information on the current status in each of these, gives examples of what has been accomplished, and suggests prospects for improvements.

RAND's Science and Technology Policy Institute was asked to provide the White House Office of Science and Technology Policy with an analysis of local law enforcement agency technology needs. This report should not only interest those in the fields of policing and criminology, but also those in the general public troubled by violent crimes and interested in steps being taken to combat crime.

Originally created by Congress in 1991 as the Critical Technologies Institute and renamed in 1998, the Science and Technology Policy Institute is a federally funded research and development center sponsored by the National Science Foundation and managed by RAND. The Institute's mission is to help improve public policy by conducting objective, independent research and analysis on policy issues that involve science and technology. To this end, the Institute

- Supports the Office of Science and Technology Policy and other Executive Branch agencies, offices, and councils

- Helps science and technology decisionmakers understand the likely consequences of their decisions and choose among alternative policies

- Helps improve understanding in both the public and private sectors of the ways in which science and technology can better serve national objectives.

Science and Technology Policy Institute research focuses on problems of science and technology policy that involve multiple agencies. In carrying out its mission, the Institute consults broadly with representatives from private industry, institutions of higher education, and other nonprofit institutions.

Inquiries regarding the Science and Technology Policy Institute may be directed to:

Bruce Don, Ph.D.
Director, Science and Technology Policy Institute
RAND
1333 H Street NW
Washington, DC 20005
Phone: (202) 296-5000
Web: *http://www.rand.org/centers/stpi*
E-mail: *stpi@rand.org*

CONTENTS

TABLES

About 95 percent of a typical law enforcement agency's budget is dedicated to personnel. The scarce resources left over are spent on basic equipment, such as cars, radios, and side arms. There is little money available to purchase the new tools necessary to keep up with criminals. Better efforts to get technology onto the streets is needed to provide modern crime-fighting technologies to the nation's local law enforcement agencies.

The Clinton administration is proposing an increase in federal assistance to state and local law enforcement agencies, to augment their resources to develop, test, integrate, and train in the use of new technological tools needed to fight crime and improve public safety.

Although crime rates have declined over the past several years, the public remains fearful and expects its government—at all levels—to do more. Law enforcement is principally a state and local responsibility; yet many jurisdictions lack the revenue base to meet the demands of technological modernization. Additionally, some specialized or expensive technologies needed only occasionally by any one local agency can more economically be provided by technology assistance from the federal level.

This report provides contextual information bearing on four facets of law enforcement technology proposals by the Clinton administration:

1. technology assistance

2. technology deployment

3. crime lab modernization

4. training.

State and local law enforcement agencies commonly work with and receive technical assistance from a number of federal agencies, including the Justice Department, the Federal Bureau of Investigation, the Federal Emergency Management Agency, and others. In addition to this, the National Institute of Justice has established a system of four regional National Law Enforcement and Corrections Technology Centers (NLECTCs), four NLECTC Special Offices, and a national center to provide what we are calling "technology assistance" and assistance with technology deployment. Each of these is collocated with technology research and development organizations, such as The Aerospace Corporation, to leverage this federal investment with existing infrastructure.

These NLECTCs have responded to some 10,000 requests for assistance annually. Examples of technology assistance include

*   audio enhancement of tape recordings

*   still-photo enhancement of surveillance videotapes

*   computer file analysis

*   metallurgical analysis

*   assistance with crime mapping analysis.

The proposed initiative would add 10 regional NLECTCs to the 4 currently operating. This would reduce the areas served by each regional center from the current 10–15 states to a more manageable 3–5 states. The intent is to provide better quick-response service to local agencies.

The NLECTC Special Offices are currently facilitating technology deployment to state and local law enforcement agencies through grants awarded for technology development, testing and evaluation of new technology, and support for technology acquisition. Technology under development ranges from a "smart gun" designed to safeguard officers from being shot by their own weapons, to safer means of handling vehicle pursuits, to a host of counterterrorism technologies. One of these offices tests and evaluates technologies by using an annual mock prison riot.

In addition, although modern crime labs are an essential part of law enforcement and criminal prosecution, the nation's laboratories vary widely in capability and capacity. The proposed initiative would help state and local laboratories modernize to better meet the need.

Local law enforcement officials have also consistently identified training as a major shortfall. Smaller departments, in particular, find it difficult to break away personnel to get the training they need. This cuts across all areas, including crime labs. The proposed initiative would build on a Department of Justice/Department of Defense partnership to make training more accessible through increased use of modern training technology, as well as National Guard and other existing training infrastructure.

Certain well-publicized instances of police being out-gunned by criminals have heightened awareness of the need to help local agencies acquire better weapons and protective technology. This report recommends federal funding in each of these areas, as well as more in-depth study to determine more definitively what law enforcement technologies are currently in use across the nation, how well they are performing, and how the federal government might most effectively and efficiently assist technological modernization.

ACKNOWLEDGMENTS

This report has benefited from comments and suggestions offered by John Ballantyne, David Boyd, Michael Camp, Barry Fisher, Kenneth Furton, Robert Greenberg, Michael Grosoman, Steven Schubert, George Sensabaugh, George Tita, and William Valdez.

Chapter One

INTRODUCTION

The local law enforcement community does not have the resources or ability to develop, test, and integrate the new technological tools that its agencies need. Training needs are going unmet, the forensics community is significantly underresourced, and the full resources of the federal government—in particular, the national laboratories—have not been turned toward the needs of local law enforcement agencies. This comes at a time when law enforcement is becoming increasingly complex and dangerous because many criminals have access to greater firepower and are more disposed to use it than the police are.[1]

In 1994, the Clinton administration developed a program to help fight crime by putting 100,000 new police officers on the street. This significantly strengthened the "thin blue line" and enabled communities to commit extra police resources in areas, such as community-oriented policing, that previously had been neglected.

The administration also recognized that law enforcement was not taking full advantage of technology and infrastructure developed for other purposes—at the cost of billions of dollars of investment—but which could also be applied to law enforcement. Accordingly, an agreement between the Departments of Justice and Defense was signed in 1994 initiating a five-year pilot program designed to in-

---

[1]The best known instance of this is the "North Hollywood shoot-out," which occurred on the morning of February 28, 1997. Outgunned by two bank robbers wearing body armor, Los Angeles police officers had to commandeer higher-power weapons from a local gun shop. Eleven officers and six civilians were wounded in the exchange—during which more than 2,000 rounds were fired.

crease the capacity of state and local law enforcement to fight crime through the application of technology.

The President's Budget for Fiscal Year 2000 proposes the new 21st Century Policing Initiative, which would, in part, help state and local law enforcement agencies tap into new technologies to share information and communicate more effectively, to solve more crimes, and to conduct comprehensive crime analysis. Specific proposals include the following:

- *Information Sharing.* Includes planning grants to states, technical assistance, and demonstration programs under the Public Safety Telecommunications Assistance Program; planning grants to states, technical assistance, and other expenses necessary to develop the Global Criminal Justice Information Network; the National Institute of Justice (NIJ) Law Enforcement and Corrections Technology Centers (NLECTCs); and other police communications improvements.

- *Crime Solving.* Includes grants, technical assistance, and other expenses to state and local forensic labs to reduce the DNA sample backlog; DNA technology research and development; improvements to the general forensic science capabilities of state and local labs; and grants to (1) upgrade criminal history, criminal justice, and identification records systems, (2) promote compatibility and participation in federal, state, and local systems, and (3) capture information for statistical and research purposes.

- *Crime Analysis.* Includes research, technical assistance, evaluation, grants, and other expenses to utilize and improve crime solving, data sharing, and crime-forecasting technologies, including funds to promote crime mapping nationwide and to promote sophisticated crime analysis models.[2]

We believe that these initiatives will be welcomed by local law enforcement agencies.

This report provides background and contextual information on some of these proposals and what they might reasonably hope to ac-

---

[2]President's Budget for Fiscal Year 2000, Budget Appendix, p. 658.

complish.[3] We begin, in Chapter Two, with a brief overview of crime and law enforcement in the United States as well as a discussion about the need for and limits to federal assistance. Then, in Chapters Three through Six, we discuss each of four technology-related areas in which the federal government has played and can continue to play an important role: technology assistance, technology deployment, crime labs, and training. For each of these, we present information on the current state of affairs, needs for improvement, instances of past or ongoing assistance, and prospects for improvement through further application of technology. Chapter Seven presents our recommendations. Table 1 outlines the topics addressed in this report and relates aspects of the three components of the 21st Century Policing Initiative to chapters most directly focusing on them.

Table 1

Overview of This Report

| Chapter | 21st Century Policing Initiative | | |
	Information Sharing	Crime Solving	Crime Analysis
1. Introduction	-----------------	Cuts across all areas	------------------
2. Context	-----------------	Cuts across all areas	------------------
3. Technology Assistance	NLECTCs		
4. Technology Deployment			Crime mapping Analysis models
5. Crime Labs		Lab improvements DNA technology DNA sample backlog	
6. Training	-----------------	Cuts across all areas	------------------
7. Recommendations	NLECTCs	DNA and labs	Mapping and models

---

[3]At the outset we should note that, to date, there has been no comprehensive survey of which crime-fighting technologies are currently being used by state and local law enforcement agencies. Nor has there been a systematic study of what resources would be needed to help bring all agencies up to minimum essential capabilities—given today's state of the art or tomorrow's expectations. Thus, the information brought together in this report should properly be viewed as suggestive, rather than definitive.

Chapter Two

CONTEXTUAL OVERVIEW

CRIME IN THE UNITED STATES

The magnitude of the crime problem in the United States is fairly well known and will be sketched only briefly here.

The most recently published detailed crime statistics are for 1997.[1] That year, in the United States one property crime was committed on average every 3 seconds. One violent crime was committed every 19 seconds. On average, there was a burglary every 13 seconds, a motor vehicle theft every 23 seconds, a robbery every minute, a forcible rape every 5 minutes, and a murder every 29 minutes.[2]

Firearms were used in 67.8 percent of murders committed (*UCR*, Table 2.11).

An estimated $15.6 billion in property was stolen. Nearly half of this ($7 billion) resulted from thefts of motor vehicles. The overall recovery rate was 37 percent (*UCR*, p. 7).

In all, more than 13 million major crimes were reported as Crime Index offenses; that is, almost 5,000 per 100,000 inhabitants. Crime rates were highest in metropolitan areas, lowest in rural counties.

---

[1] Preliminary statistics for 1998 were released May 16, 1999. Crime continued to drop, with both violent and property crimes down 7 percent. Murders were down 8 percent, rapes 5 percent, robberies 11 percent, aggravated assaults 5 percent, burglaries 7 percent, larceny-thefts 6 percent, and motor vehicle thefts 10 percent, from 1997.

[2] *Uniform Crime Reports for the United States, 1997* (henceforth, *UCR*), Chart 2.1. Unless otherwise indicated, all crime statistics cited are for 1997.

The 1997 crime rates were the lowest since 1974. Nationally, the Crime Index rate fell 3 percent from the 1996 level, 10 percent from the 1993 level, and 13 percent from the 1988 level (*UCR*, p. 7).

There were 2.7 million arrests for Index crimes in 1997; 1.5 million of these were larceny-theft arrests. Overall, law enforcement agencies reported a 22 percent clearance rate[3] for Index crimes (*UCR*, p. 7). The violent crime clearance rate was 48 percent. The clearance rate for murders was 66 percent; for forcible rape it was 51 percent, and for robbery it was 26 percent (*UCR*, p. 12).

STATE AND LOCAL LAW ENFORCEMENT

U.S. businesses and families are protected against crime by some 18,000 state and local law enforcement agencies that employ approximately 800,000 officers. More than 90 percent of those agencies have 24 or fewer sworn officers and 50 percent have 12 or fewer officers. Yet state and local law enforcement agencies are responsible for handling 95 percent of the crimes that people care about—murders, rapes, muggings, burglaries, etc.

As President Clinton has observed, "We have to recognize . . . that most laws—criminal laws—are state laws and most criminal law enforcement is done by local police officials."[4]

Each year, the U.S. public pays on the order of $100 billion to fight crime. Although crime rates are decreasing—for reasons that are not well understood—the cost of the criminal justice system keeps increasing, more than doubling in the decade from 1982 to 1993. The federal system pays about 19 percent, with the remainder paid by state and local agencies.

More than $44 billion is spent for police protection, with state and local governments paying more than $36 billion (81.8 percent) of

---

[3]The usual way a crime is "cleared" is by arrest of one or more suspects. One arrest may "clear" a number of crimes believed by police to have been committed by the same person or persons. Crimes can be cleared other than by arrest—for example, if a suspect is killed in a shoot-out with police.

[4]President Bill Clinton, remarks before Ohio Peace Officers Training Academy, London, Ohio, February 15, 1994.

that. Judicial and legal expenses to government are more than $21 billion (78 percent paid by state and local governments), and corrections costs are more than $32 billion (with 92 percent paid at the state and local levels).[5]

Local governments pay more than $31 billion for police protection; this is about 72 percent of the national total. Localities pay more than $10 billion annually for corrections; state governments pay about twice that (*Sourcebook*, p. 3).

On average, there are 250 full-time officers per 100,000 residents, with 150 of these being local police officers (*Sourcebook*, p. 38). About half are assigned to street patrol. Only about 8 percent of these sworn officers are on the streets at any given time (Sherman, 1995, p. 329).

THE NEED FOR MODERN TECHNOLOGY

Research has found that police response speed is not the major determinant of whether an on-scene arrest takes place and whether witnesses are locatable—rather, the time it takes a citizen to report a crime is the important variable (Pate et al., 1976; Van Kirk, 1978; Spelman and Brown, 1982). Here, the question arises, How can citizen reporting be accelerated, perhaps through new technology?

The availability of certain information about a crime—such as the estimated range of time when the crime occurred, whether a witness reported the offense, whether there was an on-view report of offense, if usable fingerprints were retrieved, and if a suspect was described or named—has been shown to be a valid predictor of whether the crime will ever be solved (Greenberg et al., 1975; Greenwood et al., 1997). These "solvability factors" have been used to help decide whether to follow up on a case (Petersilia, 1987), and they can presumably be brought to bear on the problem-solving approach (Eck and Spelman, 1987) to community-oriented policing.

In the past, police information has not been organized into a uniform, formatted, and comprehensive set of cross-referenced retriev-

---

[5]*Sourcebook of Criminal Justice Statistics 1997* (henceforth, *Sourcebook*), p. 3. These data are for 1993, the most recent available.

able data files available to all officers (Bittner, 1990). There is interest in improving computerized capability to make lists, produce maps, and carry out statistical analyses, building on the Drug Market Analysis Project of NIJ (Manning, 1992). Information technology has great promise for improving crime analysis (NIJ, 1996). More generally, technology can act as a "force multiplier" (NIJ, 1995).

New technologies—including electronic surveillance, data access and retrieval, and computerized decision aids—may well improve policing, but specific benefits are commonly not known, and some seemingly useful technologies go unused when offered or fielded. New technologies bring new skill requirements and can alter social relations within an organization (Manning, 1992). For example, lack of user expertise has been an obstacle to use of mapping software for crime control and prevention (Rich, 1995). Earlier evaluations of Computer Assisted Dispatching (CAD) concluded that CAD installations did not yield promised results. Laptop computers, cellular phones, and expert systems have not been subject to published evaluations (Manning, 1992). By and large, the existence of patrol-car allocation models has not created a demand for them (Petersilia, 1987). Without training and testing, new technologies cannot be properly evaluated.

Modern urban police departments need to keep abreast of and, where appropriate, make use of the latest developments in information processing, communications, and enforcement technology. Improved technology has the potential to increase police effectiveness—provided it is wisely chosen, well implemented and integrated, and appropriately used. But it is also possible for an organization to waste money chasing after the latest whiz-bang technology, bringing change so rapidly that everyone is always having to learn a new system and nothing ever seems to work right.

Given this state of affairs, it is no wonder that many law enforcement agencies lack confidence in their ability to make good information technology decisions.

THE NEED FOR FEDERAL ASSISTANCE

Americans support a decentralized system of law enforcement because they believe it allows local control and oversight—their police and sheriff departments are responsible to them, not to some far-away authority. Yet, it is this very decentralization that has made it difficult for local law enforcement to use advanced technologies to fight crime. Decentralization poses obstacles to information sharing and works against achieving economies of scale. In addition, many municipalities face severe budget constraints.

Although crime control is primarily a state and local responsibility, the federal government clearly has a role to play. In some respects this role remains poorly developed. Areas in need of further development include (1) better collaboration with local police to interdict illicit gun traffic and (2) provision of "public goods" that states and localities need but cannot afford on their own.

> These public goods include creation and maintenance of shared operational databases such as the National Crime Information Center (NCIC); *fostering and evaluating a wide range of innovations and disseminating the results of evaluations of those innovations* so that the successful ones can be replicated elsewhere; and organizing and sponsoring the information and new insights they generate (Blumstein, 1998, pp. 18–19, emphasis added).

Increasingly, governments at all levels are being asked to provide quantitative assessments of the effectiveness of their activities. As budgets come under increasing pressure and scrutiny, government agencies are turning to technology in hopes of increasing productivity—but the contribution of technology to productivity enhancement is very difficult to measure. Developing new measures for public investment in technology generally requires an assessment of the costs and benefits within a framework of public goals for the application of technology to provide the context for the investment. The technical challenge is how to plan for technological innovation as part of continuing, sustainable improvements in both community policing and law enforcement—and how to measure the effectiveness of new technologies.

LIMITS TO FEDERAL ASSISTANCE

There are also political and administrative challenges—or limitations—to be considered. The politics of federal assistance to state and local agencies is outside the scope of this report, but we include one quotation for readers to consider:

> Almost $4 billion per year in federal crime prevention assistance is given out primarily on the basis of population rather than homicide rates. Put bluntly, the money goes where the votes are, not where the crime is (Sherman, 1998, p. 42).

Similarly we note the existence of administrative challenges that policymakers should recognize:

> [The] national government now has little ability to implement policies that depend almost entirely on state and local governments for their actual, day-to-day administration, and which aim at changing the behavior of countless people in government and in the community.
>
> The record in such areas as youth and family policy, environmental policy, welfare policy, transportation policy, and health policy shows how human and financial resources can be drained from the leaky bucket of administrative federalism.
>
> Whether the administrative barriers to an expanded national role in crime control can be overcome remains an open question. But to continue to debate and analyze crime policy without due consideration of these administrative challenges is to exaggerate the ease with which crime can be affected by public policy in general, and by national policy in particular (DiIulio et al., 1995, pp. 461–462).

Legacy of the LEAA

Given the desire of federal policymakers to do something to help state and local authorities fight crime, the great temptation is simply to "throw money" at the problem. Critics contend that is what the Law Enforcement Assistance Administration (LEAA) did, frittering away $8 billion while the crime rate continued to rise (LEAA, 1980).

Established in 1968 and disestablished in 1982, the LEAA was a vehicle for federal funding of state and local governmental crime control efforts. Basically a check-writing agency, the LEAA sponsored law enforcement training institutes for state and local officials, began to develop national criminal justice data-gathering and information-sharing networks, spurred ambitious criminal rehabilitation programs, and encouraged local community-based crime control initiatives (DiIulio et al., 1995, p. 453).

The LEAA is generally viewed as having failed to meet expectations. Its virtue in allowing thousands of recipient state and local agencies flexibility in acquiring and employing technology and other resources may also have been its failing, because it did little to make policy or administration of law enforcement more coherent across jurisdictions. The LEAA has been described as one of several "major domestic policy initiatives that stalled or sank after striking the administrative icebergs of intergovernmental implementation" (DiIulio et al., 1995, p. 457).

The response to the perceived failure of LEAA was—for several years—to cease federal funding assistance to local law enforcement, in effect "throwing the baby out with the bath water."

What's Different Now

The LEAA gave law enforcement "consumers" more "buying power." What it did not give them was anything akin to *Consumers Reports*. Current efforts remedy this by providing responsive technology assistance (the subject of Chapter Three) and by facilitating responsible technology deployment (the subject of Chapter Four).

RESPONSIVE TECHNOLOGY ASSISTANCE

The current system of NIJ crime technology centers is intended, in part, to provide state and local law enforcement agencies with *Consumer Reports*–type testing, evaluation, and technology assistance.

The existing centers were established as a relatively modest effort. Because of demand for these services, the system is overextended and cannot provide the "quick response" assistance that is needed nationwide. The four regionally based centers in the network serve 10–15 states each—too much territory to cover with existing resources.

What is proposed is to establish 10 additional NIJ crime technology centers to meet the technology assistance needs of law enforcement. These centers will be supported by a consortium of federal laboratories to ensure that the best technology and science information and the best forensic technologies are available to law enforcement officers. This consortium will in no way duplicate the work of the existing state and local crime lab system. It is intended that the consortium will provide forensic support in coordination with local crime labs only when the local labs do not have the equipment or technical skills to do the job.

CURRENT SYSTEM

The National Law Enforcement and Corrections Technology Centers are part of the National Institute of Justice Office of Science and

Technology (OS&T).[1] NLECTCs provide criminal justice (law enforcement, corrections, and court) professionals with information on technology, guidelines and standards for these technologies, objective testing data, and technical assistance to implement these technologies.

The current system of NLECTCs was designed to leverage limited federal dollars to partner with and augment existing infrastructure. The center network is intended to provide a quick-response capability to agencies that need specialized technical help. In 1997 alone, the NLECTCs responded to 10,000 local requests for assistance.

The NLECTC system includes the national center in Rockville, MD, and four regional centers operating in North Charleston, SC (Southeastern region); Denver, CO (Rocky Mountain region); El Segundo, CA (Western region); and Rome, NY (Northeastern region). Also included in the system are four special offices: the Office of Law Enforcement Standards (OLES), the Office of Law Enforcement Technology Commercialization (OLETC), the Border Research and Technology Center (BRTC), and the National Center for Forensic Sciences (NCFS).[2] Table 2 gives an overview of the present system.

Technology assistance can be thought of as being one of three types: general, specialized, and referral.

- *General assistance* available through a state or local agency's regional NLECTC. This may deal with specific law enforcement or corrections problems, as illustrated by most of the examples cited in this report. Its success depends in part of the regional centers' being well connected with and responsive to the state and local agencies they serve. Expansion of the present system would reduce the regional focus of each regional center from the present 10–15 states to a more manageable and responsive 3–5

---

[1]Descriptions of the centers are taken from the NLECTC web site (*http://www.nlectc. org*).

[2]The President's Budget for Fiscal Year 2000 has listed the technology centers to be funded under the Community Oriented Policing Services (COPS) Initiative, rather than under the Bureau of Justice Assistance (BJA), but they would continue to be administered by the Office of Justice Programs (OJP).

Table 2

National Law Enforcement and Corrections Technology Centers

Center	Budget	Focus
NLECTC-National, Rockville, MD	$2.5 million	Serves as an information clearinghouse, testing products, maintaining product lists, and answering questions about technology from law enforcement agencies. Publishes newsletters and papers describing latest advances.
NLECTC-Northeast, Rome, NY	$1.3 million	Specializes in computer, communications, and intelligence-gathering technologies. Areas of research and development include weapons detection, automatic booking system, automatic firearms identification, and computerized language translation systems.
NLECTC-Southeast, North Charleston, SC	$2.2 million	Develops technology for prisons, including information management systems and simulation technologies for training. Identifies and distributes surplus federal equipment to prisons.
NLECTC-Rocky Mountain, Denver, CO	$1.8 million	Specializes in communications, weapons research, and mapping technology to help analyze crime patterns and deploy officers. Works with Sandia National Laboratories to develop technology for detecting and defusing explosive devices.
NLECTC-West, El Segundo, CA	$1.5 million	Develops microscopes and other sensitive instruments for examining barely detectable evidence. Develops technology to analyze and enhance audio, video, and photographic evidence.
BRTC, San Diego, CA	$875,000	Specializes in strategies and technologies to control illegal immigration.
OLETC, Wheeling, WV	$2.8 million	Solicits and encourages manufacturers to commercially produce new technologies for law enforcement.
NCFS, Orlando, FL	$1.4 million	Focuses on arson and explosives research to improve methods of analyzing debris from fires and explosions.

SOURCE: Rob Gavin, "Technology Helps Catch Criminals," *The Post-Standard*, Syracuse, NY, date unknown.

state average. This can be expected to further encourage utilization by local agencies of "their" NLECTC.

- *Specialized assistance* available through one of the special offices or possibly through a regional center having the desired specialized equipment or expertise. This meets the need for high technology centers possessing specialized equipment and personnel to provide services not normally available in state or local agencies or laboratories because of their cost, the extremely specialized nature of the work, or infrequent use.[3] Specialized assistance needs to keep pace with the current state of the art and should, in fact, serve to advance it.

- *Referral assistance* for uncommon expertise available through the NLECTC and other centers and offices. In this role, the centers would act as clearinghouses to help state and local agencies find the expertise they need. Success depends on the centers' information being current, comprehensive, and readily available.

National Center

The national center of the NLECTC is located at Aspen Systems Corporation, an employee-owned, information management company.

The national center coordinates the technical information collection and dissemination program for the entire NLECTC system. In this capacity, the national center

- produces detailed test reports, user guides, and bulletins on public safety equipment, metallic handcuffs, pepper spray, and DNA profiling

- operates an equipment, technology, and research information hotline

---

[3]One example of such specialized equipment is a synchrotron light source providing an X-ray beam that can be coupled with a microscope on the beam line; such equipment would be used infrequently by any individual lab but would get sufficient use at the national level. (J. Ballantyne correspondence, March 12, 1999.)

- publishes a quarterly newsletter, *TechBeat*, and maintains the NLECTC system World Wide Web site, JUSTNET, at *http://www.nlectc.org/*

- maintains current information on the manufacturers of commercially available equipment and developers of law enforcement and corrections products and services, provides referrals to users of these products, and assists in locating experts in a particular field

- helps identify equipment and technology requirements of local, state, and federal criminal justice practitioners by coordinating conference and advisory group activities, including those of the Law Enforcement and Corrections Technology Advisory Council.

Regional Centers

Southeast Center. The South Carolina Research Authority (SCRA) serves as the host of NLECTC-Southeast. Along with SCRA and other technical partners, the center offers access to an extensive, multidiscipline technology infrastructure. Highly responsive to the needs of its regional criminal justice community, which includes federal, state, and local law enforcement and corrections agencies in a 15-state area, the center also serves agencies nationwide. The center's primary focus is the exchange of information with the criminal justice community on current and emerging technologies. In this capacity, the center

- applies analytical tools and information engineering concepts to assist in developing strategic plans to meet information technology needs

- investigates distance learning technology and simulation and training technology tools for their potential application to training requirements

- plans and coordinates corrections technology demonstration projects that assess the utility of selected technologies

- assesses the application of transportation security technologies

- conducts vulnerability assessments of regional schools and reviews lessons learned by state and local law enforcement in order to develop and assess candidate school security technologies

- develops and assesses smart-card technology needed for controlled access, personnel and equipment accountability, and recording of health care services

- helps state and local law enforcement and corrections agencies acquire excess and surplus federal property, a nontechnical area of involvement.

Rocky Mountain Center. The NLECTC-Rocky Mountain office is located on the campus of the University of Denver and is affiliated with the Denver Research Institute.

NLECTC-Rocky Mountain focuses on communications interoperability and facilitating communications among different agencies and jurisdictions.

This facility works with law enforcement agencies, private industry, and national organizations to implement projects to identify and field test new technologies to help solve interoperability problems and conducts research on ballistics, weapons technology, and information systems.

NLECTC-Rocky Mountain also houses the newly created Crime Mapping Technology Center, the training and practical application arm of NIJ's Crime Mapping Research Center, which is staffed by NIJ social scientists and scholars who utilize crime analysis research to improve police field operations and develop crime-mapping software for small, medium, and large departments.

Sandia National Laboratories has been designated as a satellite of NLECTC-Rocky Mountain. In partnership with NLECTC-Rocky Mountain, the laboratory focuses on technology for detecting and neutralizing explosive devices (Operation Albuquerque).

Western Center. NLECTC-West is housed on the grounds of The Aerospace Corporation, a nonprofit corporation that provides technical oversight and engineering expertise to the U.S. Air Force and the U.S. government on space technology and space security systems.

NLECTC-West draws on The Aerospace Corporation's depth of knowledge and scientific expertise to offer law enforcement and corrections agencies the ability to analyze and enhance audio, video, and photographic evidence.

This NLECTC facility also has available an extensive array of analytic instrumentation to aid in criminal investigations, such as a scanning electron microscope, an X-ray microscope, and a mass spectrometer, all of which are used to process trace evidence.

Other areas of expertise at NLECTC-West include

- computer architecture
- data processing
- communications systems
- technologies to stop fleeing vehicles.

Northeast Center. NLECTC-Northeast is located at the United States Air Force Research Laboratory, Information Directorate, at the Rome Research Site (formerly Rome Laboratory) on the grounds of the Griffiss Business and Technology Park.

NLECTC-Northeast sponsors research and development efforts into technologies that address command, control, communications, computers, and intelligence.

This center draws on the expertise of air force scientists and engineers in its development of technologies that can be used to detect concealed weapons on people—an effort that is expected to yield a stationary device for use in buildings and handheld devices for patrol officers. Other areas of research and development at NLECTC-Northeast include

- through-the-wall sensors
- audio and image processing
- timeline analysis
- computer forensics

- secure communications
- command/control.

Special Offices

Office of Law Enforcement Standards. OLES was established as a matrix management organization in 1971 through a Memorandum of Understanding between the U.S. Departments of Justice and Commerce and was based upon the recommendations of the President's Commission on Crime. The OLES mission is to apply science and technology to the needs of the criminal justice community, including law enforcement, corrections, forensic science, and the fire service. Although its major objective is to develop minimum performance standards, OLES also undertakes studies leading to the publication of technical reports and guides.

OLES assists law enforcement and criminal justice agencies in acquiring, on a cost-effective basis, the high-quality resources they need to do their jobs. To accomplish this, OLES

- develops methods for testing equipment performance and examining evidentiary materials
- develops standards for equipment and operating procedures
- develops standard reference materials
- performs other scientific and engineering research as required.

Since its inception, OLES has coordinated the development of nearly 200 standards, user guides, and advisor reports. Topics include handguns, soft body-armor testing, pathogen and slash-resistant gloves, and analytical procedures for developing DNA profiles.

The application of technology to enhance the efficiency and effectiveness of the criminal justice community continues to increase. The proper adoption of the products resulting from emerging technologies and the assessment of performance of equipment, systems, methodologies, etc. used by criminal justice practitioners constitute critical issues having safety and legal ramifications. The consequences of inadequate equipment performance or inadequate test methods can range from inconvenient to catastrophic. In addition, these deficiencies can adversely affect the general population when they increase public safety costs, preclude arrest, or result in evidence found to be inadmissible in court.

Office of Law Enforcement Technology Commercialization. OLETC is a program of the National Institute of Justice and is located in the National Technology Transfer Center at Wheeling Jesuit University, Wheeling, WV.

OLETC is dedicated to developing and refining new strategies to accelerate the commercialization of innovative law enforcement and corrections products. It is committed to bringing high-value, well-tested products to market in a quick and affordable manner.

OLETC's staff include experienced project and commercialization managers, engineers, technical specialists, and law enforcement professionals who are able to assist developers and manufacturers in bringing new products to market. Their services include

- matching new technologies and product concepts to specific needs
- identifying technologies to develop new or improved products
- assisting with market assessments and business plans
- locating complementary technologies, expertise, and test resources as well as identifying product standards and test protocols
- identifying investment capital financing and grant opportunities
- locating manufacturing and distribution partners
- assisting with questions on liability, intellectual property, and licensing fees
- developing innovative product acquisition strategies
- developing creative funding initiatives
- providing informational and educational videos.

Border Research and Technology Center. The BRTC executive office is located in the same building that houses the Southwest Border High Intensity Drug Trafficking Area (HIDTA) and is adjacent to the

California Border Alliance Group (CBAG)—the California/Mexico border HIDTA partnership.

The BRTC works with the Immigration and Naturalization Service, the U.S. Border Patrol, the U.S. Customs Service, the Department of Defense Counterdrug Development Program, the U.S. Coast Guard R&D program, the White House Office of National Drug Control Policy/Counterdrug Technology Assessment Center, and the U.S. Attorney for Southern California, as well as with state and local law enforcement agencies and organizations operating along the U.S. borders.

BRTC is working with these agencies and their organizations in the development and implementation of SENTRI (Secured Electronic Network for Travelers' Rapid Inspection) technology as well as human presence detection, seismic sensor upgrade demonstrations, evaluation of night vision and thermal imaging technologies, vehicle immobilization, and communications interoperability.

National Center for Forensic Science. This center is a joint project of the University of Central Florida and the National Institute of Justice. The goal of the NCFS is to create a unique laboratory facility staffed and equipped to service the forensic and law enforcement communities in the areas of fire and explosion debris. The NCFS is chartered to

- conduct fundamental research investigations to gain insights into the basic nature of fire and explosion reactions

- provide the support needed for the development of standard protocols for arson and explosion debris

- promote the use of electronic media to access and exchange forensic information

- make educational opportunities available to practicing professionals and full-time students

- partner with the forensic, law enforcement, and business communities.

EXAMPLES OF TECHNOLOGY ASSISTANCE

The assistance needed by state and local law enforcement agencies can range from narrow, engineering sorts of expert advice to broad, integrative assistance.

The centers have received many letters of appreciation from law enforcement agencies they have helped—e.g., by improving the quality of audio and video tape to facilitate suspect identification and/or conviction. Typical assistance of this type includes reducing background noise in audio recordings and extracting high-quality still photographs from videotapes. Below are examples of these and other types of assistance presently in use or needed at law enforcement and other public agencies.

Utica Arson Strike Force

The Utica, New York, Arson Strike Force was assembled in April 1997 to help combat arson.[4] Because of the severity of the problem, Utica is the only northern city to be included in the National Arson Initiative, which included the southern church fire states in 1996. The strike force consists of federal, state, and local law enforcement agencies along with state and local fire agencies, which are organized into teams to perform cause and origin investigation and analysis, and technical support. NLECTC-Northeast has provided a local area network and technical support to establish a model investigative strike force as a demonstration site for the rest of the country. Prior to April 1997, the Utica arson closure rate was 2 percent (the national average is 15 percent); it is now approaching 60 percent with a 100-percent conviction rate.

NLECTC-Northeast is continuing to monitor computer network use by the arson strike force in order to utilize it as a model for other investigative strike forces. The arson strike force currently uses a computer network for preparation of court documents, e-mail communication, and Internet access—which enhance the performance of the investigators and increase the closure rate of arson cases. The initial success of the arson strike force was so overwhelming that old cases

---

[4]This project description is taken from the NLECTC web site, *http://www.nlectc.org/*.

were reopened and have been closed by arrest. The need for lead-tracking and case-management software has been identified; it is currently being developed through the Oneida County District Attorney's office. Utica has also identified a drug problem in the city and, with assistance from NLECTC-Northeast, is modeling a drug task force after the arson strike force. In addition, one of the lessons learned from the arson strike force is that there is a need for an information-sharing infrastructure among law enforcement agencies. Because of this, a shared database concept is being developed including the necessary computer-aided-dispatch/records-management system infrastructure called the Central New York–Law Enforcement Network, which will facilitate sharing and analysis of information among agencies in four counties in central New York. It is hoped that this infrastructure will be applicable throughout the country.

The Utica Police Department has observed that

> the unanticipated outcomes of this project now equal the planned outcomes in terms of benefits. The partnership with NLECTC-NE has proven to be of equal value as the project itself. The staff of the center have devoted their time, technical expertise, experience, energies and patience to assisting us in achieving our goals. They have provided insight, information, and direction. They have acted as advocates, researchers, technical advisors and facilitators (Office of the Chief of Police, City of Utica, letter, February 27, 1998).

Technology Assistance to a Fire Department

The Los Angeles Fire Department identified the following areas where it believed NLECTC technology assistance could be of assistance (Los Angeles Fire Department, letter, December 10, 1998):

- Enhance audio and videotape recordings that are related to arson and/or fire cause determination.

- Analyze fire scene debris to identify the presence of accelerants in trace amounts to help in fire-cause determination.

- Perform the analysis and set benchmark testing criteria to substantiate the accuracy of the Arson Investigation Section Accelerant Detection Canine Team.

- Conduct computer modeling of radio systems for analysis of operational variables.

- Analyze Global Positioning System data for dispatch priority and tracking of personnel in hazardous environments.

- Perform computer modeling for emergency incident planning.

- Design information network and architecture for computer system development.

Audio Enhancement Processing Helps Build Murder Case

During the investigation of the kidnapping, rape, and strangulation murder of a young woman, investigators identified two suspects believed to be responsible for the crimes. These two suspects were placed together in an interview room and their conversation was recorded.

Because of their whispering and the background noise present, investigators could not hear a large portion of the conversation. The only portions of the recording that were understandable were self-serving statements made by the suspects, probably intended to be overheard by investigators.

NLECTC processed the tapes to lower the background noise and enhance the whispering to understandable levels. After reviewing the enhanced recordings provided by NLECTC, investigators were able to understand enough of the suspects' conversation to determine they were discussing several different versions of their alibi. With the addition of this new information, investigators were able to obtain criminal complaints charging the suspects with the murder and special circumstances enhancements, making the subjects eligible for the death penalty (Sheriff-Coroner Department, County of Orange, California, letter, July 2, 1998).

Still Photo from Videotape Prompts Confession

When the suspect in a robbery/kidnapping case was shown a photo of himself, enhanced by NLECTC from a 7-Eleven store videotape, he confessed to three robberies and the kidnapping (Los Angeles Police Department, letter, March 17, 1998).

Computer File Analysis Identifies Victims

NLECTC assisted an 11-month investigation into a mail theft and counterfeit check ring, which was operating in California and Nevada. NLECTC provided information, documents, and computer technology needed to access computer programs seized during the arrest of the suspects. These computer files identified hundreds of victims. As a result, 250 members of the ring were identified and/or arrested. Several of the suspects were involved in other serious crimes, including rape, arson, use of explosives, drug sales, and burglary. Over 8,000 items of property were booked and over 25 search warrants served. Without the center's help, investigators may not have achieved this outcome (Los Angeles Police Department, letter, September 22, 1998).

E-Mail Analysis Nabs Child Molester

The Charleston, SC, Police Department requested assistance from NLECTC-Southeast in an investigation involving a missing 13-year-old boy who was very active on the Internet and e-mail services. Working with technology partner South Carolina Research Authority, NLECTC staff were able to access e-mail messages between the missing boy and a known child molester. This lead to the successful recovery of the missing boy from the Myrtle Beach area and a subsequent FBI fugitive investigation leading to the arrest of the child molester in Virginia and his return to South Carolina for prosecution.

NLECTC Metallurgy Expertise Discredits Murder Alibi

An Oregon case involved a man suspected of killing his wife and burning their house to the ground. The defense in the case put forward several theories regarding the fire cause; these were stumbling blocks to the prosecution because of the lack of expertise available, either from the State of Oregon Crime Lab or the FBI Lab. NLECTC technical expertise in the fields of metallurgy and fire cause provided technical and scientific information and examinations of evidence that enabled the prosecution to conclusively eliminate several of the defense's theories. This was credited as a significant contribution toward the successful prosecution of the defendant (Washington County, Oregon, Sheriff's Office, letter, May 15, 1997).

Videotape Enhancement Clears Suspect

Technology assistance has also helped clear suspects who were innocent, as happened when NLECTC enhancement of videotape from a hospital proved that a suspect was visiting a hospitalized relative at the time of the carjacking/kidnapping under investigation (Los Angeles County Public Defender, letter, November 19, 1998).

Technology in Strategic Planning

The following illustrates the range of factors involved in addressing technology issues in strategic planning for a typical mid-sized, progressive police department:

- A police department has begun to install computers in all its squad cars, and it needs to know how to anticipate the use of all the data/records it will create in the future to enhance the value of their use for the department and the community.

- It is interested in validating and reexamining the measures that have traditionally been used to assess the effectiveness of policing.

- It wishes to explore ways to incorporate and integrate technology into its communications and interactions with the general community.

- It is anxious to develop a "generic blueprint" for the next 5 to 10 years to guide integration of technology into the department and to identify and anticipate the changes that will be engendered by the use of technology in every aspect of its operations: patrol effects, investigation, administration, and management; it is interested in measuring the effects on such projects as Community Oriented Policing and Alcohol Interdiction.

- It wants to anticipate the future organization of its department based on the application of technology and availability of real-time information to the community and police operations.

- It wants to prepare for the future selection, hiring, and training of officers and department personnel needed to apply technology and data access in the future.

Crime Mapping and Database Challenges

Even matters that may appear to be purely technical, such as crime mapping and database maintenance, can involve much "softer" factors, as John O'Connell has observed:

> Local analysis that goes beyond crime counting is recognized and embraced wherever it happens.
>
> The most striking example of this is computer-aided mapping of crime in communities. This single advancement of knowledge about crime in a community is causing a paradigm shift in how police recognize, deal with, and plan for crime.
>
> Getting States and local communities to do this level of crime analysis requires a change in how criminal justice analysis and planning are viewed at the State and local levels. First, a research unit needs to have the authority to work with single or multiple jurisdictions across the State.
>
> Lack of money and leadership to encourage development of State and local community crime and policy analysis are the greatest limitations to better understanding crime in our communities.
>
> DOJ's Safe Streets initiative . . . is expanding community crime analysis to include employment status, child abuse, and educational success as crucial variables to better understand crime in our communities.
>
> [C]ombining the traditional criminal justice study topics with the points of view of other disciplines, while potentially fruitful, will mean at least two technical changes. First, as we find the need to integrate criminal justice and social services databases, we will need to work through confidentiality requirements. Second, optimal analysis would allow us to commingle an individual's data from various disciplines. This will be problematic because all data systems have difficulty in positively identifying and tracking individuals. The problems of individual identification will increase significantly as we try to join databases (O'Connell, 1998, p. 95).

PROSPECTS FOR IMPROVED TECHNOLOGY ASSISTANCE

Prospects are quite good. As mentioned, each of the present NLECTCs leverages federal dollars by partnering with existing infrastructure—NLECTC-West with The Aerospace Corporation, NLECTC-Rocky Mountain with Sandia National Laboratories, and so forth. The NIJ is confident that similar infrastructure partnerships can readily be developed for the 10 additional technology centers being proposed.

Chapter Four

TECHNOLOGY DEPLOYMENT

The need to develop, test, and field new law enforcement tools remains as compelling as ever, given the rapidly increasing technological capabilities of criminals. U.S. government national laboratories that produce advanced technologies and U.S. businesses that supply those technologies to local law enforcement agencies would benefit from a focused effort to develop and deploy crime-fighting technologies to local agencies.

About 95 percent of a typical law enforcement agency's budget is dedicated to personnel. The scarce resources left over are spent on basic equipment, such as cars, radios, and side arms. There is little money available to purchase the new tools necessary to keep up with criminals. As the examples below demonstrate, a new effort to get technology onto the streets is needed to provide modern crime-fighting tools to the nation's local law enforcement agencies.

CURRENT STATUS

NIJ has funded the research, development, testing, and evaluation of numerous technologies to help law enforcement. These range from less-than-lethal technologies, to weapons detection, see-through-walls systems, capture nets, a "smart" gun, a rapid DNA identification system, and more. There are numerous cases where new DNA techniques have helped free people who were wrongly convicted. NIJ has also put on the streets the technologies that it is helping to develop. For example, police now have a better way to stop a fleeing car with less risk to officers and bystanders because of an NIJ technology called RoadSpike. The following are examples of technologies currently in use by, or being developed and tested for, law enforcement agencies.

Technology Available to Campus Police

The Bureau of Justice Statistics has compiled data on nonlethal weapons authorized for use by officers in college campus law enforcement agencies.[1] Fifty-six percent of campus police departments authorize use of pepper spray, 45 percent use of collapsible batons, 34 percent PR-24 batons, 30 percent traditional batons, 11 percent personal tear gas, 5 percent large-volume tear gas, 5 percent the carotid hold,[2] 2 percent the choke hold, 2 percent stun guns, and 1 percent flash/bang grenades.[3]

As of 1995, 99 percent of campus law enforcement agencies used some type of computer. Ninety percent used personal computers, 62 percent a mainframe, 33 percent had a local area network, 6 percent used handheld and 2 percent car-mounted mobile digital terminals (*Sourcebook*, p. 42).

A Computer System to Track Gang Members

In March 1998, California Governor Pete Wilson announced an $800,000 state investment in a computer system to track gang members. The CalGang system, first developed in Los Angeles, will enable any law enforcement agency in the state to access a master index and share information about the movement of street gangs. Agencies will be able to get on the Intranet system with computer software they can buy for less than $30. The system will give them photographs of gang members and other identifying information.

Wilson also announced expansion of a program to track more than 33,000 active prison parolees. Information on the statewide computer network includes a parolee's record, aliases, physical description, addresses, vehicles, and even identifying marks, such as tattoos.

---

[1]We have not found similar data on municipal police departments.

[2]The carotid hold applies pressure to the carotid arteries supplying blood to the head. If applied too aggressively, it can cause death.

[3]*Sourcebook*, p. 42. These are 1995 data.

Wireless Data Communications

In 1997, the Guilford, Connecticut, Police Department announced completion of its wireless data communications pilot, which allows officers to gather vital suspect information and complete paperwork while remaining in the field.

The officers are equipped with laptop computers with wireless modems, enabling them to access applications and data in Guilford; from the state's mainframe motor vehicle and warrant information in Hartford, CT; from the National Crime Information Center, in Washington, D.C., and from the National Law Enforcement Telecommunication System, in Phoenix, AZ.

Guilford officers have immediate access to information on their computer screen without having to request it by voice radio— enhancing security and leaving the voice dispatch system free for emergency calls. The portable computers have increased productivity at all points, freeing officers from needing to return to the station for information and allowing dispatchers and clerical staff to concentrate on their own tasks.

The Guilford Police Department funded the purchase of the computers and the software through the Federal COPS MORE Program. To qualify for the grant, the department had to show a concrete plan for using mobile computing to keep officers in the field and increasing their contact with the public. The federal grant funded 75 percent of the project, and Guilford contributed 25 percent.

TECHNOLOGY DEVELOPMENT

NLECTC has awarded grants for development of technology in the areas of communications, computers and software, forensics, less-than-lethal weapons, protective equipment, security equipment, surveillance and sensor equipment, vehicles and equipment, and weapons and ammunition. Each of the following development projects—and many others—are described on NLECTC's web site at *www.nlectc.org/techproj:*

- Face Recognition Technology for Internet-Based Gang Tracking
- Integrated Law Enforcement Face-Identification System

- Weapons Team Engagement Trainer
- DNA Human-Identity Testing Using Time-of-Flight Mass Spectrometry
- Forensic Investigations Information Management System
- Various less-than-lethal technologies
- Concealable Body Armor
- Back-Scatter Imaging System for Concealed Weapons Identification
- Handheld Acoustic System for Concealed Weapons Detection
- Ballistocardiogram Human Presence Detection Technology Demonstration
- Vehicle Stopping RoadSpike.

Some of these technologies are quite different from what laypeople might think of as useful to law enforcement agencies. For example, one of NLECTC's projects developed software to help law enforcement agencies make cost-effective decisions for disposal of police patrol vehicles. The software enables agencies to generate a list ranking individual vehicles for disposal to minimize fleet maintenance and repair costs.

The federal government is not the only funder of law enforcement technology development.

Contributions from the private sector should not be overlooked. One of the strategic recommendations from a study conducted by the Los Angeles Police Department was that "departments should establish active liaisons and affiliations with private sector industries engaged in the development of advanced technology, which is either already adapted to or capable of being adapted for law enforcement needs" (Paniccia, 1998, p. 10).

States and a few larger cities also fund technology development. California, for example, has made a substantial investment in a computer system to track gang members, as mentioned earlier.

NIJ's Office of Science and Technology regularly convenes committees of experts to review technological issues bearing on law enforcement and corrections. This Law Enforcement and Corrections Technology Advisory Council has identified the following 11 high-priority concerns, each of these are discussed below and in Chapter Five:

- Nonintrusive detection of concealed weapons and contraband
- Safer vehicle pursuit
- DNA testing
- Officer protection
- Less-than-lethal incapacitation
- Information management
- Counterterrorism
- Crime mapping
- Location and tracking
- Secure communications
- Noninvasive drug detection.

Smart Gun

The Smart Gun Prototype II program is a one-year project to develop a firearm that will only fire for a recognized user.[4] When energized, the gun emits a radio signal that is received by a small transponder worn by the authorized user, which returns a coded radio signal. When the gun receives the signal, a locking pin is removed from the trigger mechanism, enabling the gun to be fired.

FBI data show that about 16 percent of the officers killed in the line of duty are killed by a suspect armed with either the officer's own firearm or that of another officer. In addition, there are currently 10,000 civilian firearms-related injuries and deaths each year due to accidental discharge or unauthorized use of a firearm.

---

[4]This description is from NLECTC's web site, *www.nlectc.org/*.

The need for and applicability of this technology has been aptly described in an NIJ-funded document issued by Sandia National Laboratories, Albuquerque, New Mexico, *Smart Gun Technology Project Final Report* (available in Adobe Acrobat format, *http://www.prod. sandia.gov/cgi-bin/techlib/access-control.p1/1996/961131.pdf*). Sandia identified the law enforcement requirements for smart firearms technology and then investigated, evaluated, prioritized, and demonstrated (by proof of concept) the most promising technologies available.

Prototype I was tested and demonstrated. These tests showed that the technology can work, that all necessary electronic and mechanical components can be made to fit inside a full-size pistol, and that authorization can be made well within the time required to draw and aim. Colt Manufacturing has been funded by NIJ to design and build Prototype II, which will contain more-advanced designs suitable for the law enforcement environment.

Vehicle Stopping

Police pursuits of suspects in motor vehicles have received considerable public attention—in large part due to live coverage by local television news programs.

From 1990 to 1994, an average of 331 people nationwide per year died as a result of police pursuits; 78.9 percent of these were in the pursued vehicle. An average of about 68 "uninvolved" people died each year as a result of high-speed pursuits. In California, approximately 2 percent of all pursuits result in serious injury, and about one-half of 1 percent involved a fatality. Research conducted to date indicates strong public approval of police pursuits as long as certain safeguards are in place.[5]

The Pursuit Management Task Force recently convened by the NIJ offered the following priority listing of recommendations:

---

[5]This section quotes extensively from Bayless and Osborne (1998).

1. More resources for research, development, testing, and commercialization of viable pursuit termination, management, and prevention technologies, to include

   - an accelerated "Phase III" program to deliver electrical, electromagnetic, or other technology prototypes for operational testing and evaluation by local and regional law enforcement agencies

   - encouragement of civilian law enforcement agencies to participate with vendors and suppliers in developing prototypes

   - encouragement of continued cooperation between agencies of the Department of Defense (DoD) and civilian law enforcement to transfer appropriate defense technology for law enforcement use.

2. Aggressive development of

   - retractable direct injection electrical systems

   - radiative electrical systems, including high-power microwaves

   - cooperative systems with law enforcement activation

   - auditory/visual sensory enhancements (improved warning devices).

3. A national model for collection of pursuit statistics.

4. State legislation making fleeing from lawful detention/arrest in a motor vehicle a serious crime with significant penalties.

5. Federal efforts to further public education about pursuits.

6. Research to improve interagency tactical communications technology.

Counterterrorism

Table 3 shows the most frequently cited technology needs for combating terrorism, as identified by a 1998 study sponsored by NIJ.

Table 3

Technology Needs for Combating Terrorism

Function	Need
Apprehension and riot control	Improved nonlethal weapons
Command, control, and communications	Improved and/or more readily available, secure communications for the "beat cop" Improved interagency communications
Defense against cyberterrorism	Improved detection, forensics, and counter-measures for cyber attacks
Defense against nuclear, biological, or chemical weapons[a]	Improved means to detect and categorize NBC threats Improved, affordable protective suits
Detection, disablement, and containment of explosive devices	Improved means of explosives detection Improved robots for disarmament and disabling of explosive devices Improved containment vessels and vehicles for explosive devices
Intelligence	National intergovernmental information system with current intelligence on terrorism
Surveillance	Improved "see-through-the-wall" capability Improved electronic listening devices Improved long-range video monitoring Improved night-vision devices
Training	Improved training to combat terrorism

SOURCE: TriData Corporation (1998), p. 16.
[a]Referred to in the TriData report as weapons of mass destruction.

Communications Systems

The local law enforcement community's ability to fight crime would be significantly enhanced if personnel could communicate and exchange data in a seamless fashion. Numerous cases have been reported in the news in which a lack of interoperability has led to the escape of suspects, including the murder of police officers. Communications provides the backbone that supports local law enforcement agencies' efforts to combat crime. But in most communities, radio systems supporting police, fire and emergency medical services are overwhelmed. Worse, current frequency allocations and technologies make it impossible for adjacent agencies to talk to each other. To help address this, the Clinton administration has successfully freed up 24 MHz of new radio frequencies; however, modernizing the public safety wireless infrastructure so it can take advantage of this new spectrum will be a massive undertaking.

Most communication among law enforcement agencies is by telephone, but electronic bulletin boards and web sites are increasingly used. Police planning staffs spend significant time (on average 13 percent of their time) responding to requests from outside agencies (NIJ, 1998).

The Crime Identification Technology Act of 1998 authorizes $250,000,000 for assistance to states each year, for five years, for a broad range of crime technology activities. The act provides for system integration for criminal justice purposes to help states develop and upgrade their anticrime technology from the patchwork of existing programs, integrate law enforcement and public safety records and communications, and integrate and interface with national criminal information and public safety databases.

One of the NLECTC communications projects is providing New York State with system engineering support for a wireless communications network. Another, the State and Local Communications Interoperability Analysis project, provides background data, requirements definitions, and direct assistance support to state and local law enforcement, corrections, and public safety agencies to alleviate problems associated with differences in radio systems and operational procedures. Yet another NLECTC project is demonstrating and evaluating use of telecommunications technology to provide medical care in corrections environments.

TESTING AND EVALUATION

The proposed initiative would expand the NLECTCs' current testing and evaluation programs—including partnering with the DoD testing capabilities and Department of Energy (DoE) national laboratories—so that the local law enforcement community can be confident that the tools they receive work as advertised under realistic conditions. These testing programs will enable law enforcement to make the best investment possible in new equipment and, more importantly, save lives, for example, as shown from experience with soft body armor.[6] Of the 307 models of body armor tested in 1997, only 124—about 40 percent—passed. The Bulletproof Vest Partnership Grant Act of 1998 funds provision of armored vests for police officers. In considering the act, Congress found that, although bullet-resistant materials helped save the lives of more than 2,000 law enforcement officers between 1985 and 1994, nearly 25 percent of law enforcement officers are not issued body armor.

Technologies Tested in Mock Prison Riot

In April 1998, OLETC sponsored a second annual mock prison riot, designed to give corrections and law enforcement officials from across the country an opportunity to witness a crisis management exercise that tested and demonstrated the newest technologies. Table 4 lists some of the more than 60 technologies that were demonstrated.

OLETC Commercialization Projects

Here we briefly describe several commercialization projects and the role OLETC played in them.[7]

- **Counter Point Correctional Vest.** A low-cost, lightweight stab- and slash-resistant vest for corrections officers who face a lesser ballistic threat than most police officers. OLETC provided commercialization assistance to allow the inventor to rapidly identify the potential market for this product, establish an integrated commercialization methodology, and provide the tools to effectively initiate market entry.

---

[6]The police need for body armor has increased as criminals have become more heavily armed. Of the 687 officers killed by firearms in the decade ending in 1993, more than half were killed by relatively powerful weapons: 25.2 percent by .38 caliber handguns, 12.1 percent by .357 Magnum handguns, 9.5 percent by 9 millimeter handguns, and 7.4 percent by 12 gauge shotgun (Zawitz, 1995).

[7]Information in this section is from Office of Law Enforcement Technology Commercialization (1999).

Table 4

Technologies Tested in Mock Prison Riot

Technology	Commer-cialized	Description
Body Armor	Yes	A bulletproof vest said to be capable of stopping a nine-millimeter submachine gun, a .44 Magnum, and all lesser threats
Entanglement Net	Yes	A device that shoots a five-square-meter net to stop a fleeing suspect
Night-vision devices	Yes	A head- or hand-mountable device that allows officers to see in dark areas by greatly intensifying the available light
Pupil Measurement Device	No	Detects drug or alcohol use, providing an officer with probable cause to perform other tests
RoadSpike	Yes	A long strip that lies on the ground; when a vehicle rolls over the strip, hollow point spikes embed and slowly flatten the tires
Silent Witness	Yes	A secured audio recording device carried by an officer to monitor officer/suspect interactions; can be downloaded for incident review
Spider Alert	Yes	Personal alarm designed to track employees
Tiger Vision	No	An imaging technology using invisible infrared light to give officers greater vision in total darkness

SOURCE: *Making Corrections Technology Work for You,* LRP Publications, May 1998, pp. 6–8.

- **Explosive Ordnance Disposal Technician Training Kit.** A bomb technician training kit developed by the United States Navy Explosives Ordnance Disposal Technology Division. OLETC has assisted the navy in managing the entire commercialization process by developing a commercialization plan, performing user and market analysis, managing the competitive licensee selection process, and assisting with the development of a Cooperative Research and Development Agreement.

- **Integrated Law Enforcement Face Identification System.** A special integrated facial identification system that can screen over one million mug shots in less than two seconds. The system is currently under development by OLETC.

- **RadarVision.** A see-through-walls technology using time-modulated radio signals. OLETC has assisted the developer in introducing the technology to a number of federal market sectors.

- **RoadSpike.** A remotely activated, retractable spike barrier strip to safely conclude high-speed vehicular pursuits. OLETC's role in the commercialization process was to identify candidate U.S. manufacturers, perform market and financial evaluations, and to assist the inventor and manufacturer to conclude a formal licensing agreement.

- **Tiger Vision.** A patented, low-cost, hand-held, multipurpose low- or no-light, night imaging system that operates on a standard camcorder battery or 12-volt car battery. OLETC's role in the commercialization process consisted of performing market research and financial analyses, developing a commercialization plan, profiling and identifying candidate manufacturing partners, and assisting the developer and a candidate manufacturer to conclude an acceptable license agreement.

Technology Too Good to Be True

The *Star Trek* crew used a hand-held device to detect life forms on alien planets. Is it possible to actually have such a device, to enable police to locate hidden suspects at a safe distance?

Enter the DKL *Life*Guard, with the following advertised performance characteristics:

> The DKL *Life*Guard can localize living humans up to 600 meters away, day or night, in open ground. At shorter ranges, people can be located through concrete and steel walls, earthen barriers, inside stationary or moving vehicles, and underwater. The *Life*Guard's patented dielectrokinetic technology enables it to distinguish humans from all other living things, even a gorilla or an orangutan. It is compact, portable, and effective in most weather conditions. There are no known electronic or other countermeasures to the *Life*Guard, and it is silent, passive, and undetectable. There are three *Life*Guard models, each with its key features (World Wide

Web, *http://www.dklabs.com/Model3.htm,* as it appeared February 25, 1999).

Fortunately for police departments contemplating buying these $6,000–$14,000 devices, NLECTC and Sandia tested them. Their findings:

> The results of the March performance tests were that the device failed to perform as advertised and performed no better than random chance, despite being operated well within advertised specifications and by an operator provided by the manufacturer (Murray, 1998).

SUPPORTING TECHNOLOGY ACQUISITION

Factors in Acquisition Decisions: The Example of Less-Than-Lethal Technology

Many acquisition decisions are so complex that the need for expert help is obvious. The following description of factors related to less-than-lethal (LTL) technologies illustrates this.[8]

Behavioral and Legal Considerations. To date, law enforcement has for the most part deployed a limited range of LTL weapons. The reasons for this range from cost, effectiveness, availability, and reliability issues to the unpredictability of effects and concerns about liability. For example, although a given LTL weapon may have minimal medical implications for normal healthy subjects, a high percentage of the individuals on whom law enforcement officers may use these weapons are mentally impaired or may be under the influence of drugs or alcohol. For some individuals with mental illness or those on depressants or stimulants, a technology's effects may be enhanced and longer-lasting than anticipated. Other individuals may have a higher pain threshold or impaired judgment. In addition, law enforcement and corrections may be reluctant to deploy a new technology, munitions, or LTL weapon if there are concerns about its getting into the wrong hands.

---

[8]Factors provided by Lieutenant Michael Grossman of the Los Angeles Sheriff's Department.

Training. Police training tends to focus primarily on technique, with little integration of the full range of force options an officer may employ (e.g., physical constraint) in lieu of or in combination with a given technology (Pilant, 1993). For some technologies, "one size fits all" training may be appropriate; for others, training must be tailored to the user. Some LTL weapons, for example, may require that an officer not only be tactically skilled, but also knowledgeable in the full gamut of use of force options—from vocalization to grappling to the application of lethal force. Those in specialized units, such as SWAT (Special Weapons and Tactics) teams, will have more training on movement and tactical situations and a greater exposure to a wider range of situations; whereas patrol officers may be given more-standardized training.

Cost. The adoption of a new technology must take into account the cost of equipping an entire department or specialized units, along with training costs. In law enforcement, limited budgets make acquisition and training costs a central issue. Or cost considerations might lead a department to make certain technologies available only to specialized units or designated supervisors.

Risk Management. Risk management, in practice, largely deals with damage limitation and legal liability, since most of the risk-management activity in this country takes place in an adversarial environment (Morgan, 1981). It is not enough to analyze risks or even to develop a strategy to manage risk; attention must also be given to communicating what is known about the risk to decisionmakers and the public (Leiss, 1996). Legal scholars have contributed to the field of risk communication, seeking to develop a neutral framework for characterizing the weight of evidence underlying risk assessments (Walker, 1996).

Sweetman (1987) lists the following desirable characteristics of LTL and other tactical technologies:

- Quick deployability and decisiveness in their application with a high probability of instantaneous control over a suspect

- Effects that are temporary and observable, with minimal medical consequences and predictable duration

- A high probability of affecting only the intended target

- Features that minimize the potential for abuse

- Reliability and durability

- Compactness, light weight, and ease of access

- Wide public and departmental acceptability

- Compatibility with the other force options available to an officer for a given type of situation.

An LA Story: Differences Technology Can Make

A 1989 workload study conducted by the Los Angeles Police Department (LAPD) found that officers spent as much as 40 percent of their time performing administrative duties. Patrol officers and detectives were hampered by archaic manual reporting procedures such as filling out paper-and-pencil reports, following complex procedures for bookings, waiting in line to check out vehicles or change watch, and manually filing/searching for records. Each time they arrested a suspect, they had difficulty accessing vital crime history information. Mobilization in the face of a natural disaster or other emergency was accomplished by "calling around" to see who was available for deployment. Daily field deployment relied on bulletin boards that utilized thumbtacks and magnets. The LAPD was functioning in the technological equivalent of the 1960s and had not been able to benefit from the technological advances that have increased productivity and efficiency throughout the business world.

To remedy the situation, the Mayor of Los Angeles formed a public-private alliance (called the Mayor's Alliance) that provided 1,200 computer workstations and related technology for the LAPD. Individual stations are equipped with electronic mail, voice mail, and local area network (LAN) capabilities. The LANs automate crime reporting and investigative procedures and, in the future, will include and maintain critically needed databases for personnel deployment, evidence tracking, and vehicle management. This new technology is expected to decrease the time required by officers to perform administrative and reporting tasks by 25–30 percent.

COPS MORE funding complements the Mayor's Alliance effort to create a comprehensive equipment and technology package that includes the extension of LANs to the field and to detectives assigned

to both specialized and geographic areas,[9] creation of a paperless reporting system with laptop computers, development of a data architecture for reengineering business processes within the LAPD,[10] extension of fiber-optic connectivity, installation of video case filing
(Bellow, 1993), and development of a conditions-of-probation system.[11]

Reducing the time officers spend in administrative functions will increase their productivity by an estimated 18–19 percent, which is the
equivalent of deploying 682 currently sworn officers into the field.

PROSPECTS FOR IMPROVED TECHNOLOGY DEPLOYMENT

It should be evident from the discussion above that the needs for improved law enforcement technology are great and that much new
technology is being developed. The need for testing and communication of test results is also evident. This service can most economically be provided by the federal government and, by helping state
and local law enforcement make good acquisition decisions, this
federal investment can more than pay for itself in savings to local
agencies. In addition, by promoting commercialization of technologies with law enforcement potential, better technology can become
available sooner and often at lower cost. In the next chapter, we discuss how federal funds can be used to improve the technologies in
state and local crime labs.

---

[9]Other cities adopting such systems have found that officers can save 40–60 percent of
the time they spend in preparing reports, with an overall productivity improvement of
15–30 percent.

[10]Developing a data architecture and reengineering the LAPD's outdated business
processes are expected to result in dramatic improvements in performance depart-
mentwide. Reengineering experts estimate increases in efficiency of 30–50 percent.

[11]In March 1997, the New York City Department of Probation began installing a state-
of-the-art electronic reporting system to allow probation officers to better manage
large caseloads and focus on rehabilitation and other important duties.

Chapter Five

21st CENTURY CRIME LABS

Well-equipped and well-managed state and local crime labs are a critical element of a local agency's ability to solve crimes. However, although advances are being made in forensic technology, the nation's 350 crime labs often do not have the resources to acquire and apply these technologies.

The $75 million initiative proposed by the President in FY 2000 is a big step toward dramatically improving the nation's state and local crime labs, which provide 95 percent of the forensics support to local law enforcement agencies. This initiative would establish a regional forensic support network of forensics experts from federal national laboratories, universities, and other research programs.

A crucial part of this initiative is to improve DNA testing by complementing ongoing FBI initiatives to enhance their effectiveness. This would include funds for research on rapid high-throughput technologies and methods for rapid processing of evidence samples and DNA testing, which is performed on samples both to build a felon database and to help solve crimes.

The National Commission on the Future of DNA Evidence has strongly recommended rapid development of the DNA database. In Virginia alone, this database has already identified dozens of suspects in old, unsolved cases for which there were no leads. Nationwide, over the past decade, on the basis of DNA testing, 53 people have been released from death row.[1] In Pennsylvania, as of December 1998, only about 1,000 of the roughly 10,000 samples from convicts collected had been analyzed.[2] Until this backlog can be addressed, the database cannot be fully effective, and more crimes will be committed by suspects who would otherwise be identified by DNA evidence from earlier crimes.

Arguably of greater importance is the need to provide resources sufficient to test all evidentiary samples that might help solve known crimes, including not only homicides and rapes but assaults and burglaries. As one forensics expert has put it, "Money should be assigned to solve this problem since every woman or child sexually assaulted deserves a genetic diagnosis" (Ballantyne, 1999).

The United Kingdom has been more aggressive than the United States in promoting DNA analysis,[3] which is done in the UK not only in murder and rape cases but also in burglaries—clearing hundreds of cases per week.[4] In the United States, all 50 states authorize DNA sampling of convicted felons, and every state collects samples from sex offenders. Some states, such as Virginia, collect data from a wide variety of felons; Pennsylvania takes samples only from those convicted of certain crimes (Henson, 1998). New York, following the practice of many states, allows DNA testing of certain convicts; however, New York City Police Commissioner Howard Safir has proposed taking DNA samples along with fingerprints of everyone arrested. So far, Louisiana is the only state that tests everyone arrested.[5]

---

[1]"Fla. Justice Has 'Grave Doubts' on Guilt of Some Convicts," *Washington Post,* December 25, 1998.

[2]Pennsylvania expects to add about 4,000 new samples each year (Henson, 1998).

[3]The UK is introducing many innovations through its National Strategy for Police Information Systems; some of them may be applicable to the United States. The British established the world's first national DNA database, introduced the Police National Network to link all police forces with a modern data network, and has implemented Phoenix, a national computerized criminal record database giving police and courts access to details of millions of criminal records.

[4]Although concerns about protecting privacy and civil rights may be stronger in the United States than in Great Britain, a critical review of the British experience could be useful.

[5]"Official Wants Sample of DNA Taken at Arrests," *Houston Chronicle*, December 15, 1998.

The U.S. system is neither fully automated nor fully integrated. The FBI has been successful in building consensus in the United States to use the STR (short tandem repeats) methodology, which is several orders of magnitude faster and more readily automated than the restriction fragment length polymorphism (RFLP) technology. To take full advantage of the STR methodology, local labs would require technology and training, including training of new personnel and continuing education for personnel experienced in older technologies. The training called for is comparable to college courses, which could best be provided by federal funding for FBI centralized training and for regional forensic sciences courses at local universities.

ONGOING FBI DNA INITIATIVES

The DNA Identification Act of 1994 authorizes the FBI to establish DNA indexes for persons convicted of crimes, samples recovered from crime scenes, and samples recovered from unidentified human remains. All 50 states have passed legislation requiring convicted offenders to provide samples for DNA databasing. The states have collected approximately 600,000 DNA samples and analyzed more than 250,000. All 50 states have been invited to participate in the FBI's National DNA Index System (NDIS). NDIS allows states to exchange DNA profiles and perform interstate comparisons of DNA profiles. For interested states, the FBI provides Combined DNA Index System (CODIS) software, together with installation, training, and user support, free of charge to state and local law enforcement laboratories performing DNA analysis.[6] The FBI also provides quality assurance standards for DNA testing.[7]

Of the original funding, $25 million went to the FBI and about $40 million to state and local agencies. Within four years, the FBI expects millions of genetic profiles to be available for rapid retrieval and comparison (Henson, 1998).

Already, the DNA database is credited with helping to solve nearly 200 crimes. In more than 200 other cases, "the national computer

---

[6]FBI press release, October 13, 1998.

[7]See *http://www.fbi.gov/lab/report/dnatest.htm* for a listing of these standards.

system was able to link DNA from a crime scene in one jurisdiction with DNA from a crime scene elsewhere" (Henson, 1998).

TECHNOLOGY PLAYS ROLE IN NEW ORLEANS CRIME REDUCTION

The Integrated Ballistic Identification System (IBIS)[8] at the new New Orleans Police Department (NOPD) Crime Lab is being used to prevent murder and other violent crime. IBIS can automatically link a particular gun to multiple crimes in which it was used. Each gun fired leaves unique marking on bullets and shell casings, and this evidence—once input into a central computer—allows investigators to match weapons to suspected criminals. IBIS also provides the NOPD with access to the forensic and technological expertise of the Bureau of Alcohol, Tobacco and Firearms (ATF). The NOPD now has online access to the ATF's firearms-related database, which greatly expands the ability to search for violent criminals.

During a recent 18-month period, the NOPD made 59 cold matches and numerous suspect matches between shootings and/or firearms recovered. This overall crime gun program in New Orleans is now viewed as a model for the rest of the United States to follow.

STATUS AND NEEDS

The NIJ recently published a review of the status and needs of forensic sciences in the United States (NIJ, 1999). Here we summarize findings from three topic areas: (1) training, (2) technology transfer, and (3) methods research, development, testing, and evaluation (RDT&E).

Training

Initial training currently is largely on the job. Uniform or consensus entry-level academic background requirements do not exist for all forensic disciplines. Every forensic scientist should undergo training

---

[8]The ATF's IBIS and the FBI's Drugfire system are competing bullet and cartridge case identification systems.

in quality assurance, but the number of formal classes is limited. Forensic scientists must be able to provide effective expert testimony; some expert witness training is currently available. The NIJ review's recommendations for training included the following:

- NIJ should fund forensic academic research and development programs.

- All forensic scientists should have formal quality assurance and expert witness training.

- Laboratory directors and supervisors should be provided management training.

- The profession should utilize existing and explore other delivery systems for forensic science training; examples include the FBI's LABNET and NLECTC's JUSTNET.

- Distance-learning centers accessible to forensic laboratories should be identified.

- All training needs should be funded using a combination of direct federal funds, fines and forfeitures, and foundation grants (NIJ, 1999, pp. 8–15).

Technology Transfer

The Federal Laboratory Consortium defines technology transfer as "the process by which existing knowledge, facilities, or capabilities developed under federal research and development funding are utilized to fulfill public and private needs" (NIJ, 1999, p. 17). The NIJ review (NIJ, 1999, pp. 24–26) made the following recommendations for improving technology transfer:

- Prepare a directory of technologies available at the national laboratories.

- Prepare a directory of key contacts in the national laboratories and the forensic community.

- Establish a steering committee and technical advisory focus group.

- Form strategic (working) partnerships.

- Identify sources of funding.

Methods RDT&E

The majority of crime labs in the United States engage in the following nine types of activities: [9]

- *Latent print examinations.* Although courts have for many years accepted the work performed in latent print (fingerprint) examinations, current needs include improved recovery and visualization methods, interoperability and improvement of search and retrieval systems, and shared databases for use in training and harmonization efforts.

- *Questioned document examinations.* This discipline is said to be in a chaotic state, because courts have questioned the scientific basis of handwriting identification, as well as because of ongoing changes in the ways that documents are created and transmitted. Current needs include validation of the scientific basis for handwriting examination, harmonization of comparison criteria, improved nondestructive methods for determining characteristic features of documents, image-enhancement methods for linking documents to machines, and shared databases of writing and machine-document exemplars for use in training and harmonization efforts.

- *Firearms/toolmarks and other impression evidence examinations.* Courts routinely accept identifications of firearms, tools, and other implements through comparison of microscopic impressions on questioned and authenticated specimens. Nevertheless, current needs include validation of the basis for impression evidence identifications, development of portable nondestructive analytical approaches for characterizing features of bullet impact areas, and statistical analysis of performance of algorithms used in automated pattern recognition.

---

[9]The information in this section is from NIJ (1999), pp. 27–51, which includes considerable additional discussion and explanation.

- *Crime scene response and related examinations.* The quality of analyses depends heavily on the quality of evidence recognition, documentation, collection, and preservation. Current needs in this area include sample location, identification, capture, and stabilization technology in a kit suitable for recovery of trace evidence, portable and remote hazardous materials detectors, and computerized crime scene mapping supported by the Global Positioning System (GPS) and multimedia capture technologies.

- *Energetic materials (explosives and fire debris) examinations.* See also Hannum and Parmeter, 1998. Very few laboratories routinely analyze postblast debris. Needs include improved methods for assessing the size, construction, and composition of improvised explosive devices from macroeffects at postblast scenes, enhanced cleanup techniques for postblast debris, method development for recovery of explosive and ignitable liquid residues from a variety of matrices, enhanced field-detection capabilities and mapping technologies for bomb scene investigation assistance, and continued validation of the current methods by intralaboratory studies.

- *Postmortem toxicology and human performance testing.* Although courts routinely accept these laboratory determinations, interpretive controversies still exist in several areas of toxicology. Current needs include nondestructive analytical techniques, well-controlled studies of the effects of drugs on the operation of motor vehicles and complex equipment, more-accurate methods for determining time of death, and a central database of postmortem "incidental" drug findings in deaths unrelated to drugs.

- *Forensic biology and molecular biochemistry.* Forensic DNA analysis allows for the biologic comparison between an individual's genetic makeup and biological evidence found at a crime scene. Current needs include robotic methods to replace the time consuming process of extracting biologic fluids and tissues, including differentials for semen strains; access to microchip technology to enhance and advance DNA testing methods; and sampling devices for stabilizing evidence during in-field collection.

- *Transfer (trace) evidence evaluation.* Trace evidence materials include transfer evidence of all types except biological fluids. These commonly include paints, hairs, fibers, glass, and building materials. Current needs include standardization of trace analysis methodologies, enhancements of nondestructive techniques for analysis of materials, and development and coordination of databases.

- *Controlled substance examinations.* The determination of controlled substances is the most common service delivered by forensic laboratories all over the world. Current needs include standardization of methods, automation of sampling and analysis, remote sensing equipment, and nondisruptive ("through the packaging") sampling.

EXAMPLES OF LAB CASELOADS

A recent audit (California State Auditor, 1998) of California's 19 local forensic laboratories[10] noted the following:

- "The Sacramento crime lab received 468 sexual assault cases for examination in calendar year 1997. Of those 468 cases, 317 were submitted without an identified suspect, while the remaining 151 cases had identified a suspect. Based on previous historical trends, approximately 328 of these sexual assault cases (70 percent) have potential serological evidence that can benefit from forensic DNA analysis."[11]

- Laboratories that provide services for law enforcement agencies may also support other agencies. The Kern County, California, forensic laboratory, for example, performs DNA analyses for the county's Family Support Division. It handles approximately 200 forensic DNA cases and 400 paternity cases per year.[12]

---

[10]County district attorneys' offices, county sheriffs' offices, or city police departments operate 19 local laboratories that serve the approximately 77 percent of California's population residing in 13 counties. The State Department of Justice operates an additional 11 laboratories that serve the remaining counties in the state; the recent audit covered only the 19 local forensic laboratories.

[11]District Attorney, Sacramento County, letter, December 2, 1998, included in California State Auditor, 1998, p. R-40.

[12]District Attorney, Kern County, letter, December 2, 1998, included in California State Auditor, 1998, p. R-23.

In September 1998, the San Francisco police lab "halted routine testing of DNA samples, saying the two full-time criminalists [i.e., forensic scientists] could not keep up with the work."[13]

A less exotic—but no less important—backlog exists in fingerprint identification. In some departments, there is a substantial backlog of prints taken in homicide investigations, to say nothing of lesser crimes. Delays in fingerprint identification, in effect, may discourage investigators from taking prints in the first place. This is both a staffing and technology problem. In February 1997, it was disclosed that the FBI had fallen at least three months behind in conducting background checks on suspects of murder, rape, robbery, and other felonies. With 50,000 new requests arriving every day, a backlog of 2.8 million orders had developed.[14]

Virginia's four regional crime labs review 75,000 cases annually—half being drug analyses. The labs prioritize cases by court date.[15] On average, evidence waited about 48 days in the regional labs before being tested.[16]

Given insufficient resources to meet the demand, Wisconsin forensic laboratories have had to develop case management coping strategies. Some were considered reasonable (e.g., no drug cases submitted until after all plea bargaining is over) while others were compromises necessary to cut costs (e.g., selecting one sample with the highest probability of containing semen in a sexual assault case).[17]

Yet another backlog is in the test firing and ballistics cross checking of guns suspected of being used in crimes. There are presently two

---

[13]"Attorneys Assail Police DNA Labs," *San Francisco Chronicle*, December 23, 1998.

[14]To address this, the FBI is developing an Integrated Automated Fingerprint Identification System, which will provide 10-print, latent print, subject search, criminal history request services, document submission, and image request services to FBI service provider and to federal, state, and local law enforcement users. For more information, see web site *http://www.fbi.gov/programs/iafis/iafis.html*.

[15]This is consistent with the goal of not having to go to trial without having completed lab analysis of evidence, but it may not be the best way *to solve the most cases*. Prompt analysis of the most recent evidence would likely be the most useful method for solving cases, while leads are still hot and tracking a suspect is most feasible.

[16]*Potomac News*, May 18, 1998.

[17]Michael J. Camp correspondence, March 14, 1999.

competing systems of ballistics testing: the FBI's, which is less costly, and the ATF's,[18] which is said to be better in some respects. The National Institute of Standards and Technology has brokered discussions to share databases, but further reconciliation is needed.

EQUIPMENT AND FACILITY NEEDS

> [The California audit] determined that 10 of the 19 laboratories use outmoded equipment that they must soon replace. Outmoded equipment can result in high maintenance costs, unreliable test results, and unacceptably low laboratory performance. It can also limit opportunities for staff to develop their skills using modern techniques (California State Auditor, 1998, p. 24).

The California audit noted the following (California State Auditor, 1998, p. 24):

- At least four laboratories should replace outdated gas chromatograph/mass spectrometer (GC/MS) instruments. The GC/MS is a powerful tool used to identify drug samples, arson evidence, and other materials collected at a crime scene. Replacement of the GC/MS in one lab would cost an estimated $70,000.

- At least three labs should replace their Fourier Transform Infrared Spectrometers (FTIRs). The FTIR is an expensive instrument that analyzes different substances, such as drugs, plastics, and paints. Replacement of the FTIR in one lab would cost an estimated $61,000.

- Eight GC instruments, which separate mixtures into individual components, were found in need of replacement, at an estimated cost of $35,000 each.

- Many outdated microscopes were found in need of replacement, as at estimated cost of $2,000 to $10,000 each.

The California audit estimated that it would cost more than $221 million to construct new facilities for the laboratories that do not

---

[18]In 1997, the ATF laboratories processed 2,915 forensic cases; spent 320 days providing expert testimony in the courts; spent 226 days at crime scenes; and spent 238 days providing training to federal, state, and local investigators and examiners.

currently meet the standards recommended by forensic laboratory design literature. The audit estimated an additional cost of nearly $2 million annually for the 13 laboratories without adequate quality control systems to implement and maintain these systems. It also recommended staff training that would cost an estimated $600,000 annually. The audit recommended that consideration be given to consolidating or regionalizing services, including DNA testing (California State Auditor, 1998, p. 42).

PROSPECTS FOR IMPROVED CRIME LABS

The above demonstrates the importance of crime labs and shows why funding for them to acquire modern technology and the training necessary to maintain high quality standards should be politically popular. Nevertheless, budgeting for forensics laboratories is generally hit-or-miss and is often inadequate to meet public and law enforcement expectations. Clearly, the federal government has a role to play in supporting this public good.

Courts are putting pressure on forensic science to revise Rule 702 of the Federal Rules of Evidence, to add information on the reliability of evidence, such as DNA evidence. This motivates the need for federal funding of research on the reliability of forensic evidence.

At a minimum, the federal government should encourage the nationwide development and maintenance of crime lab capabilities that are adequate to ensure that no violent felony investigation or prosecution is endangered or abandoned by lack of timely forensic support.

Chapter Six

BRIDGING THE TRAINING GAP

Local law enforcement officials have consistently identified increased training as their number one priority because they face situations that are much more challenging and complex than ever before. Yet the composition of law enforcement agencies makes it difficult for them to receive the kind of training they require. A 20-person office cannot afford to send its officers to a week of training without compromising crime-fighting capabilities, but states require continuing certification of officers in new skill areas.

A partnership between the Departments of Justice and Defense to develop dual-use training centers throughout the United States could provide local law enforcement agencies with access to interactive computer-based training and advanced simulation training tools in a location closer to home, such as at National Guard Armories[1] or other agreed-upon sites. These could help solve the problem of providing local officers with access to needed advanced training.

CURRENT STATUS

When we speak of training, it may be helpful to see specifically how wide a range of topics may be relevant. For example, the International Association of Chiefs of Police currently offers the following

---

[1]The Army National Guard has units in 2,700 communities in all 50 states, the District of Columbia, Guam, Puerto Rico, and the Virgin Islands. The Air National Guard has more than 170 installations nationwide.

training courses (see *http://www.theiacp.org/training/programs/
training_titles.html*):

Leadership

- Police Leadership: Managing the Future
- Leadership and Quality Policing
- Value-Centered Leadership: A Workshop on Ethics, Values and Integrity
- Effective Media Practices for the Law Enforcement Executive
- High Performance Work Teams

Community Involvement

- Implementing Community-Oriented Policing
- Workshop on Problem-Solving: The Seven A's
- Cultural Awareness: Train-the-Trainer
- Community Policing in America's Schools
- Civil Remedies for Nuisance Abatement
- Elderly Service Officer: Train-the-Trainer

Management and Supervision

- Managing Contemporary Policing Strategies
- First-Line Supervision
- Performance Appraisal
- Planning, Designing and Constructing Police Facilities
- Managing the Property and Evidence Function

Crisis Management

- Response to Chemical, Biological and Nuclear Terrorism

- Critical Incident Management
- Multi-Agency Incident Management for Law Enforcement and Fire Service Commanders and Supervisors

Force Management and Integrity Issues

- Investigation of Incidents of Excessive/Deadly Force by Police
- Internal Affairs: Legal and Operational Issues
- Less-Lethal Weapons Instructor Certification Course
- Less-Lethal Force Options: Concepts and Considerations in the De-Escalation Philosophy

Staffing, Personnel, and Legal Issues

- Conference on Assessment Centers and Selection Issues
- Police Law and Legal Issues: What Every Police Manager Needs to Know About the Law
- Determining Patrol Staffing, Deployment and Scheduling
- Staffing and Scheduling for Communications/Dispatch Centers
- Grant Writing for Law Enforcement Agencies
- Career Development
- Conducting Effective Employment Interviews for Entry-Level Positions
- Mentoring for the Retention of Women and Minority Public Safety Personnel

Patrol Operations and Tactical Responses

- Veteran Officer Tactical Review Course
- Patrol Response to Tactical Confrontations
- Rapid Deployment to High-Risk Incidents
- SWAT I: Basic Tactical Operations and High-Risk Warrant Service

- SWAT II: Advanced Tactical and Hostage Rescue Operations
- SWAT Supervisors' Tactics and Management
- Advanced Tactical Management for Commanders and Supervisors
- Executive and Dignitary Protection

Investigations

- Criminal Investigative Techniques II
- Managing Criminal Investigations
- Interview and Interrogation Techniques
- Investigation of Computer Crime
- Facilities Design Guidelines Survey

PROPOSED TRAINING SUPPORT

The proposed initiative would develop a partnership with NIJ, as the Department of Justice lead, the DoD Advanced Distributed Learning Initiative, and the National Guard[2] to develop dual-use training centers throughout the United States. These centers, which would be located at National Guard Armories or other agreed-upon sites, will provide local law enforcement agencies with access to interactive computer-based training and advanced simulation training tools in a location close to home. This will help solve the problem of providing local officers with affordable access to needed advanced training.

Rather than give state or local law enforcement agencies either block grants or targeted direct funding for training, the proposed initiative would focus on providing the training infrastructure that is locally available and affordable for agencies.

The main goals of the proposed training support could be to make modern, well-equipped training facilities accessible to all law en-

---

[2]One of the three missions of the National Guard is to give support to local community needs.

forcement agencies nationwide within some reasonable distance, say 25 miles, from their own duty stations. Other goals would be to develop computer-based training simulations, coordinated with other training materials, and to make them available at these partnering facilities and/or at the agencies' own facilities.

PROSPECTS FOR IMPROVED TRAINING

Advances in training simulations and other technologies offer great promise for making training more readily available. Federal support can help increase the level of training readiness, thereby contributing to public safety. Training should address specific, recognized needs, such as the need for training in crime-scene response and documentation. In addition, issues regarding training should be addressed in collaboration with leaders and visionaries in the field.

Chapter Seven

RECOMMENDATIONS

Regarding the technological needs of state and local law enforcement agencies, informed decisions should take into account an overall assessment of what technologies are in use across the nation and how well or poorly they are presently performing. For the most part, that information is lacking.

Thus, it would be useful to hold federal hearings or commission studies to determine what is happening around the country, as input to policy planning. A well-conceived study would probably more than pay for itself in cost savings and improved public safety. Such a study would probably best include both a broadly based survey of agencies across the nation and more-detailed operationally oriented site visits to a representative sample of agencies. It could provide a sound basis for establishing realistic performance goals for each element of the proposed initiative.

Presently lacking the comprehensive, empirical data we would like, this report has provided a number of anecdotal examples suggestive of broader problems and solutions.

In summary, our findings and recommendations are these:

- The *technology assistance* provided by the existing NLECTCs and their partnering organizations appears to be paying high returns on investment, helping law enforcement agencies solve crimes and protect both the public and the police. Although we cannot rigorously quantify it, there appears to be considerably more latent demand for this assistance than can currently be supplied. The pilot effort has proved itself and should be continued on an expanded scale. It would seem sensible and efficient to collocate new NLECTC sites with national laboratories or universities.

- *Technology deployment* efforts—including technology development, testing, commercialization, and acquisition support—appear to be a much needed and efficient use of federal monies. There may be a role for Department of Energy labs in advancing more-fundamental research efforts. Many technological advances in law enforcement are currently on the drawing board or in development. The need for technology deployment support will continue and will grow.

- Modern *crime labs* are essential to solving crimes, supporting successful prosecutions, and clearing innocent suspects. If the California Auditor's report is indicative of the condition of labs across the country—and we believe it is—then there are substantial needs for improved facilities, modern equipment, continuing training, and quality control. Accreditation by a central accrediting organization, as is conducted by the American Society of Crime Laboratory Directors/Laboratory Accreditation Board, is a useful adjunct to technology modernization—and sometimes drives modernization.[1]

- *Training* is commonly short-changed in the budgeting process. It is not uncommon for modern equipment to be bought and either to sit unused or misused for lack of trained operators. Failure to provide adequate training is a false economy. The federal government has a role to play in making law enforcement training more accessible and affordable.

More study is needed, but that should not delay making prudent investments, such as are being proposed by the crime-fighting technology initiative.

_____

[1]About 50–60 percent of the nation's crime labs have gone through the accrediting process. The accrediting process for crime labs looks at operating procedures but not at adequacy of equipment or budgets. It is a step toward more-comprehensive recommended standards and practices.

Ballantyne, J., correspondence, March 12, 1999.

Bayloos, Kenneth L., and Robert Osborne, *Pursuit Management Task Force Report*, El Segundo, Calif.: The Aerospace Corporation/ National Law Enforcement and Corrections Technology Center, 1998.

Bellow, G., "Video Teleconferencing," *The FBI Law Enforcement Bulletin*, August 1993.

Bittner, E., *Aspects of Police Work*, Boston, Mass.: Northeastern University Press, 1990.

Blumstein, Alfred, "The Context of Recent Changes in Crime Rates," in National Institute of Justice and Executive Office for Weed and Seed, *What Can the Federal Government Do to Decrease Crime and Revitalize Communities?* Washington, D.C., 1998, pp. 18–19.

California State Auditor, *Forensic Laboratories: Many Face Challenges Beyond Accreditation to Assure the Highest Quality Services*, Sacramento, December 1998.

Clinton, Bill, remarks before Ohio Peace Officers Training Academy, London, Ohio, February 15, 1994.

DiIulio, John J., Jr., Steven K. Smith, and Aaron J. Saiger, "The Federal Role in Crime Control," in James Q. Wilson and Joan Petersilia (eds.), *Crime*, San Francisco: ICS Press, 1995, pp. 461–462.

Eck, John E., and William Spelman, *Problem Solving: Problem-Oriented Policing in Newport News*, Washington, D.C.: Police Executive Research Forum, 1987.

Greenberg, Bernard, et al., *Felony Investigation Decision Model—An Analysis of Investigative Elements of Information*, Menlo Park, Calif.: Stanford Research Institute, 1975.

Greenwood, Peter W., J. Chaiken, and J. Petersilia, *The Criminal Investigation Process*, Lexington, Mass.: D. C. Heath and Company, 1977.

Hannum, David W., and John E. Parmeter, *Survey of Commercially Available Explosive Detection Technologies and Equipment*, Denver, Colo.: National Law Enforcement and Corrections Technology Center/Sandia National Laboratories, September 1998.

Henson, Rich, "Pa. to Join DNA Database for Aid in Finding Violent Offenders," *Philadelphia Inquirer*, December 29, 1998.

Law Enforcement Assistance Administration (LEAA), *LEAA Eleventh Annual Report, Fiscal Year 1979*, Washington, D.C.: United States Department of Justice, 1980.

Leiss, W., "Three Phases in the Evolution of Risk Communication Practice," *Annals of the American Academy of Political and Social Science*, Vol. 54, May 1996, pp. 85–94.

Manning, Peter K., "Technological and Material Resource Issues," in Larry T. Hoover (ed.), *Police Management Issues & Perspectives*, Washington, D.C.: Police Executive Research Forum, 1992, pp. 251–280.

Morgan, M. Granger, "Choosing and Managing Technology-Induced Risk," *IEEE Spectrum*, Vol. 18, No. 12, December 1981, pp. 53–60.

Murray, Dale W., *Physical Examination of the DKL LifeGuard Model 3*, Denver, Colo.: Sandia National Laboratories, October 30, 1998.

National Institute of Justice, *Law Enforcement Technology for the 21st Century: Conference Report May 15–17, 1995*, Washington, D.C.: National Institute of Justice, Office of Science and Technology, 1995.

National Institute of Justice, *Technology Solutions for Public Safety: Conference Report April 9–11, 1996*, Washington, D.C.: National Institute of Justice, Office of Science and Technology, 1996.

National Institute of Justice, "Informal Information Sharing Among Police Agencies," *Research Preview*, December 1998.

National Institute of Justice, *Forensic Sciences: Review of Status and Needs*, Gaithersburg, Md.: Office of Law Enforcement Standards, 1999.

O'Connell, John P., "Community Crime Analysis," in National Institute of Justice and Executive Office for Weed and Seed, *What Can the Federal Government Do to Decrease Crime and Revitalize Communities?* Washington, D.C., 1998.

Office of Law Enforcement Technology Commercialization, *OLETC Commercialization Projects*, Wheeling, W. Va., January 19, 1999.

Paniccia, Val, "The Impact of Informational Technology on The Los Angeles Police Department Patrol Function" in *Peace Officers Association of Los Angeles County*, November, 1998, p. 10.

Pate, Antony M., et al., *Police Response Time: Its Determinants and Effects*, Washington, D.C.: Police Foundation, 1976.

Petersilia, Joan, *The Influence of Criminal Justice Research*, Santa Monica, Calif.: RAND, R-3516, 1987.

Pilant, L. "Selecting Nonlethal Weapons," *The Police Chief*, May 1993, pp. 45–55.

Rich, Thomas F., "The Use of Computerized Mapping in Crime Control and Prevention Programs," *Research in Action*, Washington, D.C.: National Institute of Justice, July 1995.

Sherman, Lawrence W., "Cooling the Hot Spots of Homicide: A Plan for Action," in National Institute of Justice and Executive Office for Weed and Seed, *What Can the Federal Government Do to Decrease Crime and Revitalize Communities?* Washington, D.C., 1998, p. 42.

———, "The Police," in James Q. Wilson and Joan Petersilia, eds., *Crime*, San Francisco: ICS Press, 1995.

Sourcebook of Criminal Justice Statistics 1997, Washington, D.C.: U.S. Department of Justice, Bureau of Justice Statistics.

Spelman, William, and Dale K. Brown, *Calling the Police,* Washington, D.C.: Police Executive Research Forum, 1982.

Sweetman, S., "Report on the Attorney General's Conference on Less Than Lethal Weapons," *NIJ/Issues and Practices,* March 1987, pp. 14–16.

TriData Corporation, *Inventory of State and Local Law Enforcement Technology Needs to Combat Terrorism,* Arlington, Va., 1998.

Uniform Crime Reports for the United States, 1997, Federal Bureau of Investigation, Criminal Justice Information Services Division.

Van Kirk, Marvin, *Response Time Analysis, Executive Summary,* Washington, D.C.: National Institute of Law Enforcement and Criminal Justice, Law Enforcement Assistance Administration, 1978.

Walker, V. R., "Risk Characterization and the Weight of Evidence: Adapting Gatekeeping Concepts from the Courts," *Risk Analysis,* Vol. 16, No. 6, December 1996, pp. 793–799.

Zawitz, Marianne W., "Guns Used in Crime," Washington, D.C.: Bureau of Justice Statistics, July 1995.